Arthur Storer's World

Family, Medicine and Astronomy in Seventeenth Century Lincolnshire and Maryland

by

Ruth Crook

Printing Funded by Grantham Civic Society

ISBN 978-0-9929978-0-9

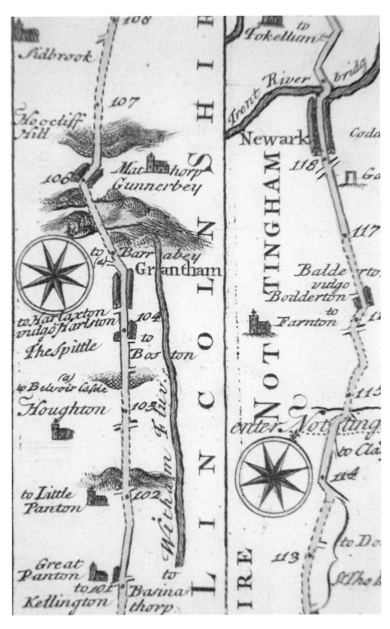

Map of part of the Great North Road, 1725

iii

Contents

Font in Buckminster church where Arthur Storer was baptised

List of Illustrations

Venus and Jupiter over Grantham Grammar School

1

Grantham and its region before 1625

Grantham is situated 110 miles north of London on the Great
North Road, which historically was the main thoroughfare
between London and York. There has been a settlement there
since prehistoric times, situated as it is at the junction of the
river Witham and the Mowbeck. Bronze Age beakers and
burials have been found, and there are later Iron Age
settlements in the area. There are also several Romano-British
settlements and one of the main Roman roads, Ermine Street,
passes within two miles. The Anglo-Saxon settlement of
Grantham developed between AD400-600. Anglo-Saxon rule
was interrupted during the ninth and tenth centuries by Danish
or Viking invasions and settlement, as shown in some of the
early street names, such as Westgate, Castlegate, Walkergate
and Swinegate. Local village names ending in 'by' are also
characteristically of Danish origin. It is likely that Christianity
spread to the Grantham area in the seventh century, but there
was a church in the town from Anglo-Saxon times. The
present day St Wulfram's church has some evidence of early
Anglo-Saxon stonework.

By the eleventh century, Grantham was a thriving market town
with a population of over 1000. Although the Domesday Book
of 1086 records the existence of the church of St. Wulfram, a
larger stone church was built in 1170-1180.[1] Grantham became
a wealthy town and, during the twelfth to sixteenth centuries,
local men amassed large fortunes selling wool.[2] Other trades
developed too, including wine sellers, butchers, carpenters,

apothecaries, weavers and dyers. When Queen Eleanor of Castile, wife of King Edward I, died at Harby in Nottinghamshire in 1290, her body lay in Grantham for one night en route to her burial in London. To mark this, an Eleanor Cross was erected on St Peter's Hill in the town.

The town contains some old stone buildings. The Angel Inn dates from the middle of the fifteenth century and may have been on the site of a much earlier building. The George Inn dates from at least 1456, although it was rebuilt in 1780. Some of Grantham House also dates from the 15th century. The Grantham Grammar School was in existence by 1327, when schoolmaster Walter Pigot was appointed as schoolmaster from 1329. The school building and headmaster's house, which are still extant, date from about 1497. Many of the other medieval buildings no longer exist, most of the older buildings dating from the 1700s, although the earlier market cross and conduit still stand in the market place. Grantham House and a timber-framed house, now known as The Blue Pig, also still stand.

Grantham House

The Blue Pig

The Angel Inn

2

Grantham and its Region under Charles I

Soon after King Charles I succeeded to the throne in 1625, a dispute began between King and Parliament, because of their differing views on the basis of his monarchy. Charles believed himself to be divinely ordained and that he could rule directly by use of his royal prerogative and without reference to Parliament, which did not meet for a decade after 1629. He was increasingly viewed by many as a tyrannical monarch, and the conflict eventually led to the civil war between his supporters and those of Parliament.

Charles had married the French Catholic princess Henrietta Maria, and associated himself with High Church divines such as William Laud, whom he appointed as Archbishop of Canterbury in 1633. Laud opposed radical reforms wanted by Puritans, and many of the king's subjects thought that he had brought the Church of England too close to the Roman Catholic Church. In an attempt to crush opposition to Anglicanism, many Puritans were persecuted, so in the 1630s some left England in search of religious freedom, most heading for New England, especially Massachusetts, in North America.

In Grantham, such differences of religious opinion led to violence in 1627. The altar in the parish church of St Wulfram's was placed in the main body of the church in an east-west direction. The vicar, Peter Titley, considered that people were not treating it with true reverence, since they were leaning on, or sitting against it. Without asking permission from the town Alderman or church officials, he had it moved to the east end of the chancel, placed in a north-south direction, and surrounded it with a rail. This reflected his belief that the altar should be removed from close proximity to the congregation and treated with sanctity. Many of the townspeople thought that this was too close to the practices of Roman Catholicism. Thomas Wicliff, the Alderman, went to the church with sergeants at mace, constables and other citizens, to reinstate the altar to its original place. They were met with opposition, fighting ensued, and the Alderman and his party were chased from the church. The community of Grantham was divided over religious doctrine and practice.[3]

Early seventeenth-century England was riven not only with religious fervour, but also with superstition. In 1613, at Belvoir Castle in Leicestershire, two of the sixth earl of Rutland's sons died following a fever, which was characterised by convulsions and sweating. A local woman, Joan Flower, and her daughters Margaret and Philippa, were accused of killing the children by witchcraft. The earl's tomb includes carvings of his two small sons, each clutching a skull. The accompanying inscription says 'both which dyed in their infancy by wicked practise & sorcerye'. Joan Flower died on the way to Lincoln to be tried for witchcraft, but her daughters were found guilty at the assizes and hanged. Many people believed in witchcraft and thought that witches were also non-conformists, as they did not attend Anglican church services.[4]

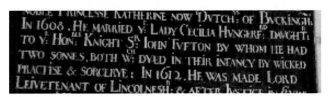

Text from the earl of Rutland's tomb

Tombs of the earl's sons in Bottesford Church

Woolsthorpe Manor, the boyhood home of Isaac Newton, had witches marks inside the doorway, to protect the family from evil spirits. It was believed by some that evil influences or witches could also enter the house and cast spells on them whilst they slept.

Witch mark by the window at Woolsthorpe Manor

Like other English towns, Grantham was subjected to outbreaks of plague, which occurred in 1604, 1617, 1625 and 1637. In the latter year, it was recorded that 68 people were buried in the churchyard in the last three weeks of May. Plague victims and people with contagious diseases would be sent to the Pest House on Manthorpe Road, until they either recovered or died. Medical treatments generally were very rudimentary and most people died if they became seriously ill. People looked for cures in many different ways. At Wysall in neighbouring Nottinghamshire, a case was brought to the Archdeaconry Court of Nottinghamshire in 1623, concerning a number of people who resorted to the 'Stroking Boy of Wysall.' This boy was credited with the power to effect cures of ailments by the action of stroking the affected area, and some twenty people were presented who admitted the charge. Witnesses were called and cases were cited of people who had travelled from Wollaton, Trowell and West Bridgford in the county to be cured by the stroking boy. One father, Richard Garton of West Bridgford, was questioned 'for carryinge his childe to Wysall to be cured by the stroaking boye.' Robert Shaw of Trowell was also brought before the same court for 'consorting with a wisard'.[5]

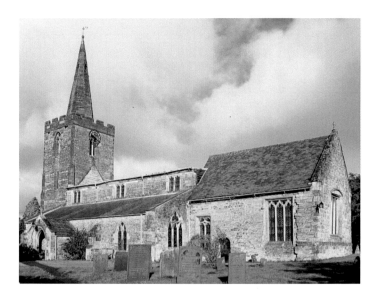

Wysall Church

Superstition and belief in magic were still commonplace in the late 1700s, when a doctor, called John Parkins, established himself as a professional cunning-man in the 'Temple of Wisdom' in Little Gonerby, now part of Grantham. He built up a considerable clientele from all over the country for his charms and un-bewitching services. He printed pamphlets and booklets to advertise, and was very popular for removing spells and providing charms for everything from prosperity to fertility.[6] At this time apothecaries and physicians treated the person as a whole rather than their illness.

3

William Clarke, Apothecary of Grantham, and the Civil War

William Clarke was born in Grantham in 1609 and baptised at St Wulfram's Church on 23 April.[7] His father Ralph, and his grandfather Arthur, were apothecaries in the town. Ralph was well respected and served as Alderman (elected leader of the council) in 1611 and 1621. William was part of a big family, and shortly after his eighth birthday, his mother died. His father remarried, and with his step-mother Cassandra had four more children, starting in 1620 with Benjamin, followed by Rachel, Joseph and Deborah. The family lived in the rooms above the apothecary shop, situated on the High Street, immediately to the north of The George Inn, where the Great North Road passed through the town. Many travellers must have passed by the shop, and William would have become used to meeting many different people. Sick or injured people were brought there to be treated, strips of cloth and equipment for treating injuries being mentioned in the shop's inventory, taken after Ralph Clarke's death. William probably attended Grantham Grammar school, as did his younger brothers. His father taught him the apothecary and alchemy trade, with which he would have become acquainted from an early age.

When he was 21, his father died, and as the oldest surviving son William inherited his property, including the well-stocked shop. The inventory shows that, as well as equipment, the shop had plentiful stores, including rock alum, green coppers, brimstone, bay oil, white lead, red lead and sugars. There were liquorice, rhubarb, almonds, rice, starch, pepper, prunes, raisins, currants, hops and honey, oils of cinnamon, cloves and nutmeg, as well as flowers of rosemary, rose, celandine, gilly flowers and hyacinth. Leaves and seeds included lemon pips and pine nuts, amongst others. There were also soups of lemon and raspberries, juices of dragon's blood, and body parts of crocodiles, French flies and other animals, and minerals and perfumes. The shop also contained scales, weights, measures, shelves, four mortars and pestles of brass, one stone mortar with a wooden pestle, the shop boards and other implements.

The rest of the building above the shop was substantial, and had at least two further floors and a garret. The rooms included a hall, parlour, hall chamber, middle chamber, two chambers over the shop, gallery, retiring chamber, gallery chamber, alcove, kitchen, stable, still room, little still room and other room, brewery, buttery and kitchen. In the yard were two young pigs and a pig sty. The back yard had a cow, a hay store, a barn containing wood, buckets, and lathes.[8]

Apothecary shop, from Gravity Fields Festival 2012

Alchemist from Gravity Fields Festival 2012

Seventeenth-century still for extracting oils from plant matter

11

Site of William Clarke's House

Basement of the current building showing original stone pillars and stone slabs in situ

Joseph Clarke carved his name on the inside walls of the
Headmaster's House in 1644

William's younger half-siblings were still very young when
their father died, so he was charged with responsibility for their
upbringing. Seven years later, on 2 October 1638, William
married Katherine Johnson at St Wulfram's church, and on 22
April 1640 their first son William was baptised. In the
following year William was appointed by the Grantham
Alderman's Court to be Mill Master for the Welham Street
Mill. The couple's second child Elizabeth was baptised on 13
March 1642, in tumultuous times when the country was
beginning to be divided between the king's supporters and
Parliamentarians, and in increasing danger of falling into civil
war.[9]

William Clarke was an ardent supporter of Parliament and
Puritanism rather than the King and Anglicanism, and was not
afraid to speak his mind. In June 1642, after the king had left
London for York, he wrote a letter to Lord Willoughby, the
Lord Lieutenant of Lincolnshire, giving his controversial
opinion on that matter. 'When the king went to York that if the

prince had stayed behind him, he should have been crowned king, and now that the prince was gone with the king, the Duke of York should be crowned king'. He also had a heated discussion with one of his neighbours in Grantham, asking him whether he was for King or Parliament. The man said that he was for both King and Parliament, so Clarke pressed him further for an answer and the man said that he would be for the king with all his heart.[10] Clarke was then reported to have said 'Thou hast a rotten, stinking heart within thee, for if thou wilt be for the King, thou must be for the papists'. The details and Clarke himself were sent by the deputy lieutenants of Lincolnshire to John Pym, leader of the Puritan opposition at the Houses of Parliament. There, his outspoken views were discussed in the House of Commons on 27 June. One of the many important men who were present was the future Lord Protector, Oliver Cromwell. It was decided that the matter would be referred to the Court of King's Bench, to which he was sent by the Serjeant at Arms. He was probably held in the King's Bench prison in Southwark, but two months later, in August 1642, the Civil War began and proceedings seem to have been dropped, since there is nothing in the court's records relating to his case.

Grantham House by Claude Nattes c. 1793

Welham Street Mill and Ducking Stool by Claude Nattes
c.1793

Castlegate, Grantham by Claude Nattes c. 1793

(27)

Beinge here at Lincolne, a petition was tendered to my Lord
Willughby, by William Clarke of Grantham, the consiera-
tion whereof, my Lord having referred to us, wee presently
writt a letter to the Alderman of Grantham, to attend
us, with such Examinatons as hee had formerly
taken in that busines. Which accordingly was
this day done, and finding the Informatons of such
a nature, as was not fitt, to bee passed over, wthout
the knowledge of the howse of Comons, wee have
sent upp the party in safe Custody to attend the
howse, and left direction with the Alderman of
Grantham, to send the witnesses to waite uppon
the howse, whensoever they shalbee called for.
Which wee thought fitt to Comunicate to you, desiringe
yo: to acquainte the howse with itt, to which
End, wee have sent upp his mittimus, a true Coppy
of the Informatons signed by the Alderman, and
soe remaine.

Lincolne: June: 22th
1642

Yor most affectionate
friend
Jo: Wray
W Armyne
Jo: H-tchcocke Christo: Wray
farther Jor by Tho: Hatcher

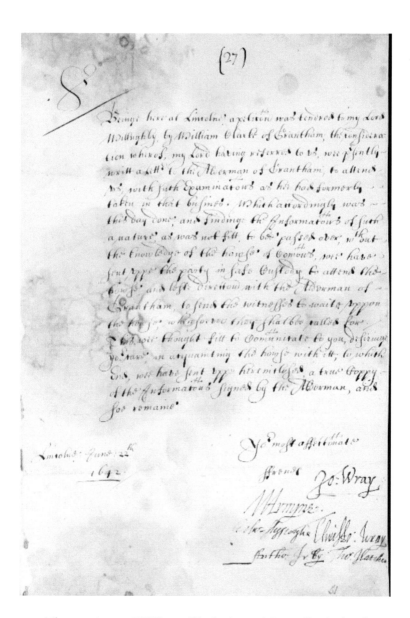

The report on William Clarke's anti-Royalist behaviour

16

The Great North Road in Grantham in the late Nineteenth
Century, The George Hotel on the left

The Great North Road in Grantham in the early Nineteenth
Century showing the Angel Inn. The old building to the right
may have been similar to William Clarke's house and shop

On 12 January 1643, Grantham, which was in a strategic position on the Great North Road, was garrisoned by a detachment of the Earl of Newcastle's Royalist army. When the troops arrived, the town was initially loyally inclined towards the King, and the newspaper *Mercurius Aulicus* reported that 'the people of the town came out to meet them with joy and music; willingly admitted two troop of horse, for the securing of that pass; and took great care to see them billeted and disposed of to their best content'. Two assaults were made on the town by the Parliamentarians in late January and early March, and then on 24 March 1643 Royalist troops stormed Grantham and took 250 prisoners to Newark. Ninety gentlemen and fifteen townsmen, including William Clarke, were charged with high treason and were due to be tried on 11 April. Grantham was taken for Parliament by Oliver Cromwell in May and 45 royalists were captured. These were exchanged for the prisoners at Newark, who were freed on 22 May. The Grantham Alderman and seven royalist comburgesses were imprisoned in Nottingham.[11] It was at that time that Parliamentarians destroyed the Eleanor Cross and removed wood from the church.

In the Grantham Corporation minutes of 1645, William is listed as a Commoner and by the end of the first Civil War in 1646 had been elected to the second twelve of the council. In the following year he was elected to the first twelve, a process which had taken him two years, but would normally have taken ten.[12] It was an important step, and a necessary pre-condition for the important role he took in the life of the town and its district during the Commonwealth and the Cromwellian ascendancy of the following decades.

William and Katherine Clarke's Trade Tokens

Apothecary jar

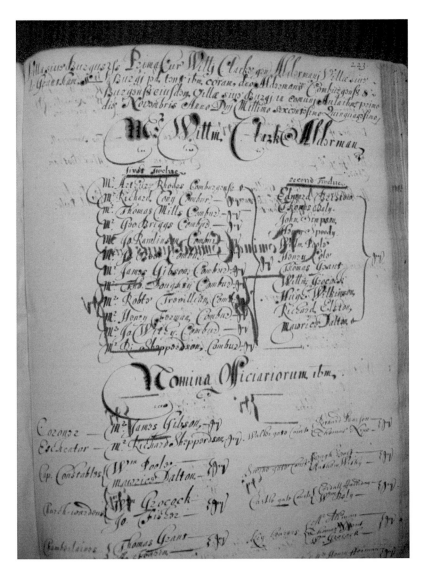

The Hall Book of Grantham entry of 1 November 1650
showing William Clarke as Alderman

4

The Storer and Babington families

William Clarke's second daughter Judith had been baptised on 16 February 1644 and his wife Katherine had died prior to 1647, when he married for the second time on 18 July that year, at Buckminster in Leicestershire. His new wife, Katherine Storer, was the widow of Edward Storer, who had been born in Buckminster in 1615 to Edward and Mary Storer. Edward's family had lived there for three generations, and he had a privileged upbringing. On 31 October 1637 he had married Marie Widmerpoole at Wysall in Nottinghamshire. Marie was the daughter of George Widmerpoole, whose family had lived at nearby Widmerpool for several generations. She may have died after the birth of their daughter Ann, and was buried in Wysall on 5 January 1640.[13]

Following Marie's death, Edward Storer married Katherine Babington, then living at Bunny Hall, Nottinghamshire, with her aunt Catherine and husband Isham Parkyns, who was lord of the manor there. Bunny was the closest village to Wysall, just over two miles away. They applied for a marriage licence on 15 February 1641, the document stating that Edward was residing in St Peter's parish in Nottingham.[14]

The Babington family had been established in Dethick, Derbyshire for generations. The main branch, although overtly Anglican, remained Catholic in private. Anthony Babington of Dethick had been involved in the 'Babington Plot', a conspiracy to release the Catholic claimant to the English throne, Mary Queen of Scots, and he was executed for treason by the government of Queen Elizabeth in 1586. Humphrey Babington, one of the younger sons of the family, moved to Rothley Temple in Leicestershire in 1567 to marry Margaret Cave, the daughter of Francis Cave, a gentleman. Their son, the Rev. Adrian Babington, an Anglican clergyman, married his cousin Margaret Cave, and they had numerous offspring, including their daughter Katherine, who was born in Cossington, Leicestershire, in 1613. Her brother, the Rev. Humphrey Babington, two years her junior, played an important part in the Storer family's life, and also that of Isaac Newton.

Edward Storer and Katherine Babington's eldest son Edward was baptised on 7 February 1642 at Bunny, before the Civil War began seven months later. Their daughter Katherine was

Bunny Hall

baptised on 8 August 1643 at Bunny.[15] It is not known if Edward Storer had a profession; in the church registers he was referred to as 'Mr Edward Storer Gent'. Following Katherine's birth, Edward, Katherine and their young family went to live in Buckminster, with the Storers' extended family. Edward died in June 1644, months before his youngest son Arthur was born. Arthur was baptised on 20 February 1645 at Buckminster. Katherine was left a widow with four small children, Anne 5, Edward 3, Katherine 1 and baby Arthur, named after his great grandfather Storer and his father's brother.[16]

Arthur Storer's baptismal record 1645

Katherine's brother Humphrey was no doubt supportive of his sister Katherine when her husband died. He was a man of many talents and interests, including music and astronomy, and had a large library, including music books, and several musical and astronomical instruments. He was educated at Cambridge University, where there were many reforming influences in the church. After obtaining his MA in 1642, he worked as a tutor at Trinity College. He was a kindly man, loved and respected by his family and friends, and was very generous to his wider family. When his cousin William King, a master weaver, wanted to start a business in London, he gave him the money to enable him to do so. Throughout their lives he played an important role.

5

Arthur Storer's early life

Arthur Storer was born at Buckminster in Leicestershire, when the Civil War in its third year. Buckminster was a Parliamentarian stronghold, with the lord of the manor Sir Edward Hartop raising a regiment for Parliament. Katherine Storer's uncle and aunt, Isham and Catherine Parkyns, were Royalists, and during the war Prince Rupert and Queen Henrietta Maria, respectively the nephew and consort of King Charles I, both stayed with them at Bunny Hall. Despite this, Katherine took the Puritan Clarke as her second husband.

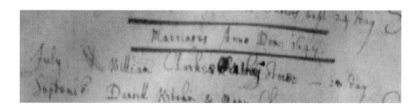

Marriage register in Buckminster showing William Clarke and Cathy Storer's wedding

On 18 July 1647, when William Clarke married Katherine Storer at Buckminster, they united two large and wealthy families. Katherine was a very attractive woman, as recounted many years later by her daughter, another Katherine. The new family, which now consisted of seven children under eight years old, went to live in William Clarke's house, above the shop in Grantham.

Buckminster church

After Parliament's victory over the king in 1646, and his election to the first twelve of the corporation in the following year, William Clarke became one of the ruling Parliamentary group in the borough of Grantham. In October 1647, all the remaining royalist supporters were removed from office in the town, following an act of Parliament of 9 September. Later he was to become perhaps the most important man in the town.

His brother Joseph Clarke who he had sent to study at Cambridge university in 1645, returned to Grantham and was appointed as usher at the Grammar School in June 1649.[17] William, a keen businessman, bought numerous properties in the town following the declaration of the Commonwealth in

1649, when many crown lands were sold off, making him one of the leading landowners by 1650. In the same year, Humphrey Babington was ejected from Cambridge University, along with many other clergy, for refusing to take the Oath of Engagement that they were obliged to take in support of Parliament. The words of the oath consisted of 'I Do declare and promise, that I will be true and faithful to the Commonwealth of England, as it is now Established, without a King or House of Lords'.[18] Babington was a fierce supporter of the King and the Royalists during the Civil War. Following his ejection from Cambridge, he spent the summer of 1651 with friends in Kings Lynn. He corresponded with his friend William Sancroft, discussing and exchanging sheet music. He liked the town and felt that it was the next best place to Cambridge.[19]

Trinity College Cambridge

In 1651, when Clarke was elected as Alderman, the leader of the town council, for the first time, he commissioned the building of a horse mill on Elmer Lane. He was one of only three Comburgesses who served as Alderman twice during the Commonwealth period, his second term being in 1657. Following his election, Henry Stokes, the master of the Grammar School, who may have been a Royalist, attended the court to ask if he was still wanted as master, and was reassured that he was.[20]

In July 1652, the spire of St Wulfram's parish church, just opposite the school, was struck by lightning and badly damaged. Scaffolding was erected and the upper part of the spire removed. Because of the troubled times, repair work could not be carried out and rain entered the steeple, rotting the bell frames. The steeple was not repaired until 1664, after the Restoration, when the Church of England had become dominant again and Clarke had been removed from his prominent position.[21]

Katherine and William had three further children, Joseph, who was born between 1648 and 1651, Martha baptised on 28 March 1652, and John, on 27 December 1653. They had many visitors to stay in their house, including acquaintances, friends and travellers. The building must have been full of bustle and noise, both inside and out. The children may have been used to travellers passing by the window and staying in their rooms. Sick and injured people were brought to Clarke, who was now a well-respected apothecary. He was also a keen teacher and nearly all his sons became apothecaries. The church records show that in May 1653 a stranger was found dead at their house.[22]

The children also met a variety of people, such as Henry More, who later became an English philosopher of the Cambridge Platonist school. More, who was five years younger than Clarke, stayed at the Clarkes' house during his visit to Grantham. He been born in the town and attended the Grammar School. His name can be seen carved on the outside of the school building.

Henry More's name on the school walls

During 1654, Humphrey Babington also went to live with his aunt Catherine and her husband Isham Parkyns, at Bunny Hall. The Parkyns remained committed royalists during the Commonwealth period. Isham offered Humphrey his churches of Keyworth and Bunny, from where he could not be ejected. In the following year he became vicar of Stanton on the Wolds.[23]

6

Isaac Newton comes to stay

When Arthur Storer's brother Edward was eight, in 1650, he may have gone to Grantham Grammar School to join his step-brother William, who was two years older. Arthur would first either have been either educated at home or attended a dame school. He may have been anxious to join his older brothers, who might have told him about the school and the master Henry Stokes and usher Joseph Clarke, their father's (and step-father's) brother. Arthur may have gone to the Grammar School in 1653, learning Latin and Greek, mathematics and religious studies. The Clarke's daughters were also educated, probably at home, and could read and write.

The children's uncle Joseph Clarke, the school usher, had a daughter Ann, baptised on 8 April 1655, who later married John Smith, a mercer from Grantham. Henry Stokes also had children, a son Samuel born in 1655, followed by Ann, baptised on 18 December 1657, Lewis on 8 November 1659 and Elizabeth on 23 December 1660. Joseph Clarke had been appointed in June 1649, Stokes in 1650.

In 1655, the 12 year old Isaac Newton came to live in the Clarke household. He boarded in the garret of the house, where Clarke's large library was kept. Newton had been sent to the Grammar School in Grantham from his home in Woolsthorpe, 8 miles away.[24]

Woolsthorpe Manor

Isaac Newton's mother Hannah Newton, like Katherine Clarke, had been widowed before he was born on Christmas Day 1642. He was a premature baby and was so small that it was said he could fit in to a quart mug. Lady Jane Pakenham of North Witham, wife of Sir Clement Pakenham, was called upon to assist, by providing medicines for the baby.

Woolsthorpe Manor in 1845

North Witham church

On 27 January 1645, Hannah married the recently widowed Rev. Barnabas Smith from North Witham, leaving the young Isaac with his grandmother Ayscough on their farm in Woolsthorpe. Barnabas was 63 at the time and Hannah was born two years after Barnabas's son Benet. Benet may have died as a child, since he is not mentioned in Rev Smith's will. Hannah's brother Rev. William Ayscough, who was rector of Burton le Coggles, was thought to have arranged the match. The name Benet was later used by both the Vincent and Storer families.

Whilst living in the small village of Woolsthorpe, the young Isaac became friendly with his neighbour Edward Storer's sons, Oliver and William, who may have been distant relations of the Storer boys with whom he later lived. They were the same age as him, so he had some company. Years later, as a tenant of Newton's, Edward would owe Newton rent for his farm.[25]

Despite living near the Storers and many cousins, Isaac Newton often spent many lonely hours by himself. He was frustrated and angry with his mother and later admitted that he had threatened to burn down his mother's house with his mother and step-father inside. In August 1653, when Isaac was 10, Barnabas Smith died, and Hannah returned home with three young children, Mary 6, Benjamin 3, and Hannah 1, step-brother and sisters for Isaac. Isaac had his mother back at long last, but had to share her with three other children. He later admitted to being peevish and thumping his sister, and being rude to the household servants.[26] He continued to attend dame schools in Skillington and Stoke Rochford.

View of Woolsthorpe from North Witham

The stream adjacent to Woolsthorpe Manor where Isaac would sail small boats

Hannah's brother the Rev William Ayscough saw potential in the young Isaac and suggested that she should send him to the local Grantham Free Grammar School, eight miles away. Hannah would have to pay a fee to send him there, because it was only free to boys from Grantham itself. Katherine and William Clarke were friends of Hannah Newton, so the Clarkes were chosen for the young Isaac to lodge with. Katherine and Hannah were also distantly related by their family's marriage into the Vincent family. Katherine's great aunt Margery Babington had married Edward Vincent in about 1540. Thomas Vincent had married Mary Newton, who was Isaac Newton senior's aunt, in 1593. Hannah must have been well aware of William Clarke's anti-Royalist and Puritan tendencies, and these views were in the ascendant in the 1650s.

Living in the town was a new experience for Isaac. Having to live with three rowdy boys, four girls and three toddlers, Isaac may have been frustrated and jealous of the other children. He later admitted to stealing cherry cobs from Edward Storer and being peevish over bread and butter.

Grantham Free Grammar School

7

School life, domestic life and local events in the 1650s

When Isaac Newton first went to the school in 1655, he was one of the less able pupils and sat on a lower form. There were 60-80 boys there, and others of his age had already had the benefit of four years of grammar school education. Like the rest, Isaac carved his name on his desk and on the school walls, as had become the custom. His name can still be seen on a window-sill in the schoolroom. After school, he tended not to go out to play with the other boys. He apparently preferred the company of the girls of the house, and made them dolls' furniture. Katherine Storer was only a few months younger than him, and they became good friends, perhaps forming a romantic attachment.[27]

Isaac became friendly with Arthur Storer, two years younger than himself, and may have walked to school with him. This was a friendship that was to last for Arthur's lifetime. Arthur may have liked the attention of an older boy and although he was younger than Isaac, he may have been physically bigger. One morning, on the way to school, it was perhaps Arthur who punched him in the abdomen. After school that night, urged on by the other boys, it may have been Arthur who fought with Isaac in the adjacent churchyard. Isaac soon had the upper hand and, encouraged by the other boys, he took hold of Arthur's ears and rubbed his nose along the church wall. A few years later, one of the things that he regretted and mentioned in his list of sins was 'beating Arthur Storer'. Nevertheless, he was to become a lifelong friend. Perhaps encouraged by his victory, Isaac progressed well with his studies and soon became the most able pupil in the school. Storer was not the only fellow pupil who he mistreated. He also said that he regretted sticking a pin in John Keys's hat to prick him. John Keys's sister Susan, later married Edward Storer.

St Wulfram's Church at night

Inside the Old School

The Headmaster's House

Garderobe in the Headmaster's House

The children who lived with William Clarke learned from him. They watched him make pills and potions, and learned what herbs and tinctures were used for various illnesses. Newton made extensive entries in his notebook on how to make pills and potions and how to treat sick people. This training led to many of Clarke's extended family becoming either physicians or apothecaries. The girls also grew herbs and learned what they were used for. William may have been the father-figure that Isaac had previously lacked, and he probably enjoyed his company. By the time Isaac stayed with the Clarkes, their elder son William may have already left school and begun working with his father, learning the profession.

During Cromwell's ascendancy in the mid-1650s, it appears that Grantham was a centre for gathering intelligence about possible royalist conspiracies in the East Midlands. An undated letter, written by a government agent about 1655, mentions that he had received intelligence from Grantham of a meeting of suspected royalists at Rufford Abbey in Nottinghamshire, the home of Sir George Savill. Various measures to apprehend those involved followed. This incident occurred during the years when William Clarke was perhaps the leading parliamentarian in the town, and it is possible that he was involved in governmental intelligence gathering activities in the area.[28]

The years before William Clarke's second term as Alderman, and his period of tenure, were marked by controversy. In October 1655, Edward Coddington was censured for uncivil speech in the court, claiming that he was uncertain whether Clarke meant to kill him or not, and saying that he was afraid of him. Coddington was fined 3s 4d for his choice of language.[29] A few weeks later when the court met again, it was voted upon and decided that the previous Alderman Edward Towne should remain senior to William Clarke.[30] In July 1656, Clarke would not give his answer to the court as to whether he would stand as Alderman later that year, as he considered that he had been wronged in his nomination and decided to evade the question. He did however relent and began his second term in the autumn.[31] By April the following year, there was controversy again, as Clarke was claiming extra benefits as Alderman, none of which had ever been claimed before.[32]

Clarke was not only vocal in attempting to persuade other people to let him do what he wanted to do, but a very exacting man too. He insisted that the Comburgesses wore their cloaks in the town, the town Constables carried their staffs, and that dogs were kept on leads.[33] Scarlet cloaks were also made for the town's musicians. In 1657, as Alderman and Justice of the Peace, he presided over marriages both in Grantham and in the Vale of Belvoir, being mentioned as doing so in the parish registers of Muston and Knipton. He was clearly also interested in technology and ordered the building of a windmill on Gonerby Hill, and supervised its building.[34] The young Isaac Newton took an immediate interest in the construction of the new windmill, and went up to Gonerby Hill after school in the evening to watch it being built. He came back to Clarke's house and made drawings of its construction, mostly on the walls of his room. He built a model windmill, which he placed on the roof of the house to enable it to catch the wind. According to Stukeley, the drawings were lost when the house was rebuilt in 1711, and the windmill had been demolished by 1727.

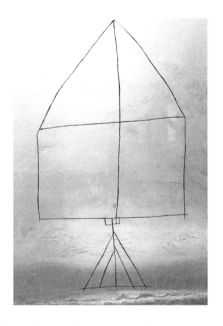

Highlighted carving of a post-mill in Woolsthorpe Manor

Post-mill very similar to the one carved in Woolsthorpe Manor

Family allegiances were also important to William Clarke. A few years earlier, when Grantham town was short of money, he lent the corporation £100, which Rev Humphrey Babington, his brother-in-law, subsequently lent him until he could be paid back. William Clarke appeared to be a man who insisted on things being done properly, and usually his way, and was exacting in everything that he did. During his second period as Alderman, he managed to persuade the Borough Court to employ Babington, a royalist, as minister of Grantham for six months. He also managed, in 1657, to secure for him tithes owed and six months pay.[35] Clarke claimed £3 from the court towards his expenses for going to London 'about Mr Babington's business'.[36] In early 1657, Babington moved from Grantham and became vicar of Easton Maudit in Northamptonshire, under the protection of Sir Henry Yelverton.

The whole family were very fond of Babington, who was a regular visitor to Grantham. He may have shared his astronomical knowledge with the boys and let them use his instruments. He gave Newton a wooden box with which he made a water clock, and which he kept at the Clarkes' house. It was about four feet high and inside was a system of wheels and pulleys and a weight that dropped as water dripped inside. The front had a dial and a needle which turned round indicating the time of day. Newton replenished the water supply each day.

In 1658, Clarke was still in arrears with rent on the mill pingle and it was agreed that he would give Babington the £3 still owed to him by the court and that the remainder of Clarke's debts should be given to the chamberlain.[37] He was clearly very careful with his money, claiming extra expenses from the court and being tardy in paying his debts.

Interior of the school in 1858

Garret room in Moseley Old Hall, Staffordshire, used for
apothecary wares

Newton enjoyed making lanterns, including one to light his
way to school, which he was able to fold up and put in his
pocket. The boys made sundials and scratch dials to tell the
time and track the motion of the stars. Isaac worked out how to
measure the force of the wind, by making marks on the wall
and measuring his jumps with and against the wind. One of
Newton's favourite books was John Bate's *Mysteries of Nature
and Art*. It described how to make kites, or 'fire drakes', with
exploding firecrackers, which the boys made and Isaac flew
over the town at night. This caused panic, as the townspeople
were frightened that it would set their houses on fire. Isaac also
confessed that he had been swimming in a kimnel on a Sunday
and squirted water, which may or may not have involved the
other children. Isaac also made himself a cart with four wheels

and a crank to propel himself around the corridors of Mr Clarke's house. The Clarkes were clearly indulgent hosts.

The boys continued at school under the watchful supervision of Mr Henry Stokes and Dr Joseph Clarke and progressed with their studies. They learned trigonometry and land surveying, as well as Latin and Greek. They may have also had the use of the many books in the Trigg Library in St Wulfram's church. As they got older, they may also have paused on their way to school to look in the window of Edward Pawlett's bookshop, which was on the main street on their route to school.

Isaac Newton's name amongst others on the window-sill of the school

Newton continued to be interested in making models and drawing on the walls in his garret. When he returned home to Woolsthorpe during the school holidays, he continued drawing and carving on the house walls. It must have seemed very quiet at his country home, after living in such a noisy environment in the town.

Apple tree through the window of Woolsthorpe Manor

Scratch dial on the outer wall of the school

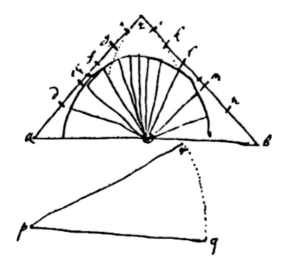

Newton's sketch of how to make a Dial for any Latitude

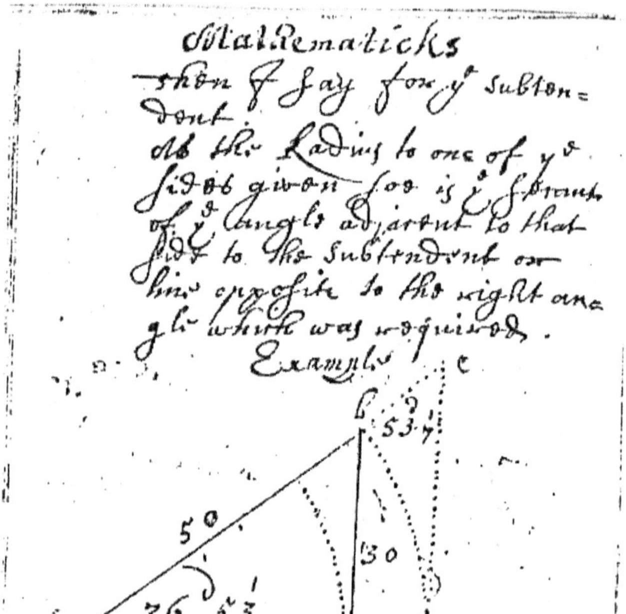

Mathematicks

then & say for ye subten-
dent

As the Radius to one of ye
sides given, so is ye sine of
ye angle adjacent to that
side to the subtendent or
line opposite to the right angle which was required.

Example.

For the Triangle (a, b, c) let
ye sides (a c) & (b c) given & ye
(b c) 30 comprehending ye
right angle (a c b) and let
the sides (a, b) be demanded
& the 2 acute angles (a b)
& (b a c) The square of (a c)
40 is 1600 & the square of
(b c) 30 is 900 which 2 squares
added together makes 2500

Page from the Mathematics book dated 1654 used at Grantham
School

48

When Isaac was 17, his mother decided that it was time for him to leave school and come home permanently to help on the farm at Woolsthorpe. He was a poor shepherd and, when he was distracted reading a book, let the sheep wander in to a neighbour's field and eat their crops. He made boats and sailed them on the stream adjacent to his house, and was clearly not a natural farmer. He had had a taste of what he could achieve and was more interested in learning. When visiting Grantham on market day, he would slip away to Mr Clarke's house to read his books and no doubt see the family. After one such excursion, he was leading his horse home whilst reading a book and failed to notice that the horse had slipped its rein and gone home ahead of him. Henry Stokes realised the potential of the boy and asked his mother to consider letting Isaac return to school if he waived the school fees. He was then allowed to finish his grammar school education and returned to board at Mr Clarke's house.

Lincolnshire Long Wool
sheep in the field
adjacent to Woolsthorpe
Manor

8

The Truman family and Maryland

During Newton's absence, in June 1660, Edward Storer went to Wadham College, Oxford, to continue his training to become a physician. At about the same time, Ann Storer left home to marry James Truman, a widower and physician at Muston in Leicestershire, six miles from Grantham.

The Truman family were staunch Non-Conformists. James and some of his brothers and sisters had been born at Sutton-in-Ashfield in Nottinghamshire, where their Independent church, founded in 1651, was the fifth oldest in the country. The main branch of the Truman family came from Gedling in Nottinghamshire, another strongly Non-Conformist community.

The House used by the Independents in Sutton in Ashfield

Gedling Church

James Truman's brother Major Thomas Truman arrived in Maryland in 1651 with his servants Symon Bird and John Totney. He helped Lord Baltimore to establish religious freedom, and in 1656 was given 1000 acres of land called Trent Neck. In the same year, he sent for three more of his servants from England, John Salt, John Hyett and Maurice. In 1657 he purchased land for his brother James, a plantation of 700 acres called Indian Creek, on the south side of the Patuxant River, where it bends to the east to Chesapeake Bay. The claim was patented in 1658. Thomas was a Burgess representing St Mary's County in the Maryland General Assembly in 1661. Their other brother Nathaniel and sister Marie had also gone to Maryland in the mid 1660s. Nathaniel owned a tobacco plantation called Truman Hills, looked after by slaves and servants, and there traded with the Indians. James clearly had plans to go to Maryland before he married Ann.[38]

Early map of Maryland, showing the Patuxent river

Ann and James Truman's daughter Elizabeth was born at
Muston on 20 April 1660, but died in the following November.
Martha was born in January 1662 and Ann in March 1664.
James Truman's sister Elizabeth, who was also living in
Muston, married Thomas Stringer in January 1663.

Muston Church

9

Grantham in the 1660s

After the Restoration of the Monarchy in 1660, the Puritan members of the Corporation were gradually removed from office. The king wrote to Thomas Grant, the Alderman, in July asking him to examine whether Richard Pearson and others had been unjustly removed from their position on the Corporation due to their loyalty to him. He asked that if this was so, 'to cause those to be put out who withheld them'.[39] On 3 October, a commission led by Sir William Thorold and Dr Thomas Hurst came to 'examine certain matters of the corporation' and William Clarke and Maurice Dalton were asked to leave and resign. In March 1661 five Royalists were restored to the Corporation by mandamus and a further four in June.

Babington gave a sermon at Lincoln assizes, published in Cambridge in 1678, which included passages of Latin, Greek, and Hebrew. He stated that 'monarchy is the best safeguard to mankind, both against the great furious bulls of tyrannical popery, and the lesser giddy cattle of schismatical presbytery'.[40] Shortly afterwards, he returned to Cambridge. In 1661 he also became Rector of Boothby Pagnell in Lincolnshire. He strongly influenced both his nephew Arthur Storer and Arthur's friend Isaac Newton during this whole period and remained close to them both for the rest of his life.

Newton finally left Grantham grammar school for Trinity College Cambridge in June 1661. It is not known where Arthur Storer worked, but he may have followed in the profession of his step-father since, in his will twenty-five years later, doctor's equipment is itemised. When Newton went to Trinity College, he stayed in Babington's rooms and is thought to have been his sizar, since his expenses show that he paid Babington's women.[41] Babington was a mentor to Newton and, in a letter to him, said that he was looking forward to reading his research.

Newton's rooms in Trinity College Cambridge

After the Isaac and Arthur left the grammar school, John Clarke may have begun to attend in 1662 when he was eight, following his brother Joseph, who was slightly older. In August that year, Joseph Clarke, William's brother, went to the Grantham Borough Court and asked if he could be relieved of his duties as usher. The court 'gave him many thanks for his care and pains bestowed amongst them as schoolmaster' and gave him permission to leave.[42] He continued his work as a physician in the town until his death 28 years later. He was buried in Grantham on 9 December 1690.

Henry Stokes also attended the Grantham Borough Court to ask if his brother Edward, also a well-educated man, could take Clarke's place as usher, which he did for a year. In 1663, it came to the court's attention that Stokes had been making secret plans to leave the school and return to Melton Mowbray Grammar School, without giving notice. The court was aggrieved about this and promptly sacked him in order to get another Master immediately. At Melton, he received an annual salary of £40, including a house free from taxes and the billeting of soldiers. His salary at Grantham had been £18 per year, out of which he had to pay the usher. He remained at Melton School until his death in 1673.[43]

In 1662 the Act of Uniformity was passed by Parliament. It prescribed that any minister who refused to conform to the form of public prayers, administration of sacraments, and other rites of the Established Church of England, as prescribed in the Book of Common Prayer, should be ejected from the church. The 'Great Ejection' followed, when two thousand Puritan ministers left their positions as clergymen. Congregationalists, Presbyterians, Baptists and Quakers were all considered to be Non-Conformists.

Matthew Sylvester, a cousin of the Storers, had been ejected from his living at Great Gonerby church near Grantham and went to stay with James Truman's cousin Joseph, who had been ejected from Cromwell church in Nottinghamshire. Joseph lived in Mansfield, where he continued his Non-Conformist ministry. Sylvester was then appointed as a private house minister to Mr White of Cotgrave, Nottinghamshire, a well known Non-Conformist, and was married to the daughter of Sir Edward Hartopp of Buckminster.[44]   Sylvester later came to know Richard Baxter, the pre-eminent Presbyterian minister among London Non-Conformists. Baxter thought very highly of Sylvester and described him in the most generous terms. Sylvester was severely impoverished and suffered illness. In 1687 Baxter, then over seventy and newly released from prison, became his unpaid assistant. Sylvester's family life does not appear to have been a happy one. In his will, written in 1704, he describes his oldest son Samuel and youngest Matthew 'to my great grief and shame have disappointed all my Cost Cares and hopes and have showed themselves incorrigibly and obstinately set against my Councills and Commands.' He left his estate to his loving wife Elizabeth and his other 'dutifull and faithfull son' Joshua. [45]

Matthew Sylvester

In 1663, although Grantham had a postmaster, letters were being intercepted and read by the Alderman. In November, Albina Vane wrote some letters in cypher to Anne Hutchinson in Grantham. The Alderman intercepted the letters and questioned Mrs Hutchinson. He had immense power within the town and three years later interogated John Pechell, a Quaker, about publishing papers. Pechell's neighbours thought him to be mad or distracted and of a deep melancholy and silent disposition, but behaved quietly.[46]

Katherine Storer left the family home in August 1665 and married Francis Bacon, a Grantham attorney. Her uncle Humphrey Babington conducted the wedding ceremony at Boothby Pagnell. Her grandson later became the headmaster of Grantham Grammar School in 1729. Newton left Cambridge in 1665 to escape the plague, and as well as returning to Woolsthorpe, stayed with Babington for a while, and so may have been present at the wedding.

Newton may have also been at Boothby with Babington on 13 October 1666, when the worst whirlwind and earthquake in anyone's memory hit Lincolnshire. In Welbourn, out of eighty stone houses in the village, only three were left standing, timber was dispersed over a wide distance and three or four people were killed. In Boothby, part of the church was blown down and many trees were torn up by the roots. At Denton, a fierce hail storm hit the village at about the same time. Hail stones up to three inches long fell, and some were said to be like darts, arrows and other odd shapes.[47]

In 1669, Babington became a Doctor of Divinity and later bursar and vice-master of Trinity College.

10

Emigration to Maryland

Ann and James Truman had emigrated to Maryland by 1669, when James was a Justice of Calvert County, and he became Commissioner in 1670. Their young daughters, Martha, Mary and Ann, also accompanied them on the long voyage, as well as nine servants. Their other daughter Elizabeth was probably left behind in England, because she was too young to travel. It is not known whether Ann's brother Arthur Storer travelled with them at this time too, but he was living in Maryland by 1672. Two of their Babington cousins also accompanied them, as did several people from Gedling, Nottinghamshire.

Ann missed her family and was a prolific letter writer. Her letter numbered 92, written on 8 April 1671 to her cousin Matthew Sylvester in Cotgrave, said that letter writing would never make up for seeing their faces. She told him that their daughter Ann (called Nanny) had died of fever during the previous year, as had one of their servants. She had also had a son earlier in the year, who had lived only for ten days. She asked to be remembered to cousin Sanderson, and remembered to all their friends in Mansfield. She thought that Maryland was a 'darke part of the world', and wrote about the practice of trading deer skins with the Indians and sending them home to England.[48]

James Truman became ill and died by 9 June 1672, aged 50, his will being witnessed by his brother in law Arthur Storer in Maryland.

I

Mr White's House in Cotgrave

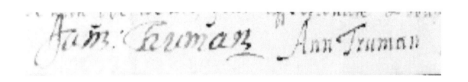

James and Ann Truman's signatures

From petuxunt river Maryland Aprill 8th 1671

Good Sr

It was noe small pleasure to us to hear of your health & welfare it being one of our greatest happyness to know our english fr. are well, & though we can not see the faces yet yt we may read the lines of each other. We have cause to bless God & thank you yt (unworthy) we are upon your heart & in your prayers. I hope we have fared better therefore: & oh yt we may be enabled by our bountifull landlord to pay to him the due rent of praise & fruitfullness & not forget him who is soe mindefull of us. Hee is pleased to visit us with various providences, our deare Nany dyed june 19th & one of our servants a litle before, the rest of us (through mercy) seasoned (as thay call it) well, most of us in the sumer (3 months of wch was hot) had an ague & feaver wch held us about a week or fortnight, but have had our healths this winter. I was delivered of a son jan. 10th who lived but 10 dayes (blessed be God) we are all well at this time. I was sorry to heare of the loss wch Coz. H: had, & though hir greif might cause hir to fall into travell about the time hir husband died, I wish there be noe scandall, (I would willingly be informed concerning it) I hope shee deserves it not. I shall be glad to hear yt you are married to hir you mentioned, who I think is a vertuous branch of very Godly parents. I am sorry to hear yt wickedness is set up by a law to the persecution of soe many good people, but glad to know yt those yt suffer for christ shall never be loosers by him, for though thay may be destitute & aflicted yet not forsaken & though we have cause to simpathize with you in your troubles yet if God would thereby transplant you into this darke corner of the world we should have reason to rejoyce. pray present our servis to mr White & his Lady & tell him we had got noe skins this year, the indians dress none till sumer, good buck-skins are usually worth 6s 8d & doe-skins about 3s 4d. I pray you present our due respects & servis to frs at Mansfeild, to coz. Sanders sons, I am glad my deare babe is with soe good a nurse. if you have oppurtunity pray present our hearty love to mr Vaughan & mr James their familys, we should be glad to hear thay are well. Deare Coz. the Lord be with you we are your affectionate & obliged fr.

Jam: Truman Ann Truman 24

my maide sends [illegible] hir servis to you

Ann and James Truman's letter to Sylvester

Ann Truman married Robert Skinner, a widower with three children, Robert, Nathaniel and Mary, not long afterwards. They went on to have three children of their own, Clarke in c.1673, William in c.1674 and Adderton in c.1677, two of whom became apothecaries.

Arthur Storer was a keen astronomer and, armed with his astrolabe, took regular readings of the stars and planets. Astronomy had been studied at Harvard from the late 1630s, but Storer was the first astronomer in North America who received international recognition. He studied the stars and planets and calculated their positions, which he made into tables. He may also have worked as an apothecary, because his will later mentioned a 'Parcell of Docters meanes'.

Some of the children whose parents had left Gedling in Nottinghamshire with the Trumans and been born in Maryland, were also recorded in the Gedling church registers.

It is unknown whether William Clarke's daughters Judith and Elizabeth lived to adulthood, as the parish registers are incomplete, but a Judith Clarke married Thomas Parnam at Grantham in November 1671, and an Elizabeth Clarke married Richard Tomson in 1676. This Judith had six children, whilst Elizabeth only had one. William Clarke and some of the members of his family had moved to Loughborough in Leicestershire, where he wrote his will in 1671, but his daughters Judith and Elizabeth were not mentioned. He added the clause at the end of his will that 'If any be not thanken that person or persons shall have none'.[49]

Major Thomas Truman, James's brother, was the subject of controversy on 26 September 1675 after massacring some

Indians on the north bank of the Piscataway Creek. The Indians had fortified themselves in the old fort of the Piscataways across the Potomac river from Mount Vernon. On 26 September Major Thomas Truman and soldiers from Virginia besieged the fort. The siege continued for seven weeks and, when the Indians' food was nearly exhausted, they sent out five of their leaders to negotiate. They asserted that they were fast friends of the English, and not the Indians who had been murdering white people in the area. In proof they exhibited a Maryland medal with its gold-and-black ribbon, which had been given them by Governor Calvert as a token of amity and a protection. Thomas Truman took them to one side as if to negotiate, then beat them on their heads and brutally killed them all. Major Truman was impeached by the lower chamber of the Maryland legislature, found guilty by the upper, fined and sentenced to a prison term. There was great revulsion against this savage act. He was dismissed from the Council in 1676, and released from his bond for good behaviour in 1678.[50]

In 1677, Nathaniel Truman died in Maryland without issue, so his land and belongings, including slaves, were left to his nieces Mary, Martha (who may also have been called Ann) and Elizabeth, and there were also small donations to their mother Ann Skinner and his two sisters, Mary Truman and Elizabeth Stringer of London.[51]

Edward Storer left university in 1664 and settled in Buckminster, where he married Susan Key, sister of his school friend John. He was a well respected physician in the area. He had six daughters, Elizabeth, Mary, Anne, Katherine and two called Susannah, who both died in infancy, and four sons, Edward, Francis, Thomas and Arthur, Arthur dying in childhood. Anne married her cousin Ellis Key and had three sons, whilst Francis had a son Bennet and a daughter Catherine.

Arthur Storer returned to England and stayed for a while with his uncle Humphrey Babington in Boothby Pagnell. On 10 August 1678, he wrote to Isaac Newton in Cambridge and included a table of astronomical calculations. The table gave the hourly altitude and azimuth of the North Star. Arthur wrote that he would desire Newton to examine the table and give him 'notice of the truth thereof as may be', for he 'shall very shortly be for Maryland: hoping to be at London by the latter end of this month or the beginning of the next September'. He added that if he should be in London before Newton wrote back, he would be staying with his cousin William King, a master weaver, at Three Horse Shoe Alley, off Old Street in London. He would then leave for Maryland from Gravesend.

Arthur wrote again to Newton on 4 September from London, beginning the letter, 'our former acquaintance imboldeneth me to trouble you'. He enclosed tables of the Sun's azimuth for each hour and also a table of the North Star's hourly azimuth. He emphasised that he was using a new method, which he had not heard of before, to calculate azimuth directly, without first finding the altitude, and was eager to ascertain Newton's opinion.

Boothby Pagnell Church

He also wrote to his uncle Babington on 19 September, telling him that William King and his wife were both well and their business was also prospering, and that they had employed a boy apprentice and a journeyman.[52] Arthur also quotes a letter that he had received from Newton in reply to his two letters. Newton said 'I have looked over your tables of the North Star's hourly altitudes and azimuth and do not perceive but they are sufficiently exact'. In Arthur's letter to his uncle, he also included some other tables and explained the reason for his calculations and methodology and that he was extending the tables to show azimuth for each degree of latitude. He also thanked Babington for 'your trouble and great kindness at Boothby and elsewhere'.

Arthur's final letter to his uncle before leaving for Maryland was dated 1 October.[53] He again mentions that he has received another letter from Newton. He says that it had encouraged him to 'present him with another bird of the same feather hatched at the said time and I think as well trimmed and decked with feathers yet not altogether of so swift a wing but fully as many changeable notes'. He encloses further tables for his uncle to show Newton, but concludes that he may not get an answer before leaving England. The 33 year old Arthur appears to be excited at the prospect of returning to Maryland, which is observed by the levity and wittiness of his letter.

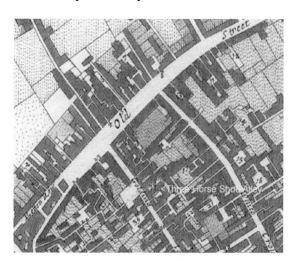

Three Horse Shoe Alley, off Old Street, London

Ship leaving from Gravesend

Cliffs marking the entrance to the Patuxent River

After several months away, no doubt his sister Anne would have been eager to see him again and learn news of the family that she missed back in England. There may also have been news of her daughter Elizabeth, left in England.

One of Arthur Storer's letters to Newton, 1678

11

Arthur Storer and the comets

Arthur wrote again to Babington on 18 April 1681 from Mr Kinder's land at Hunting Creek on the banks of the Patuxent in Maryland. He told him that, on 2 October 1680, he had fallen from his horse, causing what he suspected was an inner bruise, and that he had been very weak ever since. He said that he had not since travelled further than two miles from his house, where he lived with his sister Anne. He said that sometimes he thought he was getting better and sometimes not, and was worried that he had a consumption.[54]

He enclosed a table of the motion of the Great Comet and asked his uncle to ask Newton to look over the results. He had spent many nights observing the sky over Maryland, with his primitive astronomical instruments, charting the movements of the celestial bodies.

Newton had observed the Great Comet himself almost daily from 12 December 1680. It had disappeared from view briefly and then reappeared two weeks later, before finally disappearing during the following March. Initially, it was thought that there were two comets, but it was soon realised that they were one and the same. Newton took measurements of its tail and made a sketch of the comet over King's College Chapel, as it had been described to him by Babington.

In 1681, a library, designed by Sir Christopher Wren, was being built at Trinity College. Babington paid for two chambers above four arches on the south side of the building, which was assigned a garden surrounded by a brick wall. He had the use of the rooms for life, as did anyone bearing the name of Babington or Cave. Any other member of his family who was not a fellow of the college could use the upper room. If the rooms were rented out, the rent was to be used to purchase books for the library.

In 1682, all their friends and family were no doubt saddened to hear of the death of William Clarke, who was buried in Loughborough.

By 1683, Arthur Storer had regained his strength and was writing letters to his family and friends. He wrote again to Newton on 26 April that year, sending him measurements of the comet of 1682. He had observed it from Indian Creek on the opposite bank of the Patuxent river from his earlier observations of the previous comet, at Mr John Hunt's house near that of Colonel Joules (or Jowles). He asked Newton to send him some instruments, including a large fore staff, and some star charts. He said that he had received a letter from his brother Edward telling him of a comet that was visible in England. He also told him that he had written to his mother Katherine Clarke, his brothers Edward, John and Joseph and his uncle Humphrey. Newton still kept in contact with the family and was interested in news about them. Storer commented regarding the comet, that he must be 'one of the first that took notice thereof in Maryland'. The comet was eventually named Halley's Comet, since Edmund Halley was the first to identify its previous appearances and forecast its next. Storer also wrote about a solar eclipse on 17 January 1683 and several lunar eclipses, which he measured and charted. He mentioned 'many Noteable conjunctions' including those of Saturn and Jupiter. All these observations were later found to be very accurate, and were some of the most accurate readings of the time.[55]

Colonel Jowles was an influential man in Maryland. He had studied at Oxford University and entered Gray's Inn, although in later life was called a doctor and surgeon. His education brought him prominence in Maryland, where he came to hold many senior positions.

In 1685 the other Truman brother, Thomas, died. As he was also without issue, he left his land and possessions to his brother's daughters. The three sisters were now very wealthy and owned vast tracts of land and tobacco plantations. In 1697, Elizabeth and her husband Charles Greene, an apothecary in Kings Lynn, conveyed the inheritance to Thomas Greenfield, Martha's husband, the transactions being witnessed by Robert Sparrow, the Mayor of Kings Lynn. Elizabeth and Charles Greene had a son and two daughters, Charles, Ann and Mary.

Back at home in Grantham, Katherine Bacon married John Vincent in 1685. It was many years later, as an elderly lady, that Stukeley interviewed her about Newton.

Astronomer using a staff to measure a comet

Map of Maryland 1676 showing settlements along the Patuxent
River

12

The final chapter

Arthur Storer became ill in late 1686 and made his will in November that year. Early in 1687 he died and was buried on his sister's land at the Old Reserve, now on the site of Calvert High School, Prince Frederick. In his will, he mentioned his sisters Katherine and Ann, his brother Edward, his mother Katherine and his half-brother Joseph. He left his Universal Double Ring Dial, an astronomical instrument, to his brother Edward, and enough money to buy his mother some new gloves. He also left clothes to John Hunt, with whom he had stayed when he had observed the comet.

Double Ring Dial

Shortly after Arthur's death, Newton's famous work *Philosophiæ Naturalis Principia Mathematica*, was published on 5 July 1687. Newton acknowledged Arthur Storer in the work for all the observation and readings of the stars and the comet, a fitting epitaph for a skilled astronomer, scientist and mathemetician.[56] 'The same day, *Mr. Arthur Storer*, at the river *Patuxent*, near *Hunting Creek*, in *Maryland*, in the confines of *Virginia*, in lat. 38½°, at 5 in the morning (that is, at 10^h. at *London*), saw the comet above *Spica* 角, and very nearly joined with it, the distance between them being about ¾ of one degree'. Later he writes 'And Mr. *Storer* (by letters which have come into my hands) writes, that in the month of *December*, when the tail appeared of the greatest bulk and splendor, the head was but small, and far less than that which was seen in the month of *November* before sun-rising; and, conjecturing at the cause of the appearance, he judged it to proceed from there being a greater quantity of matter in the head at first, which was afterwards gradually spent'.

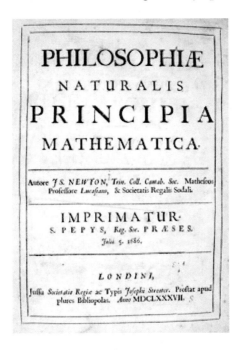

73

In 1689, the Act of Toleration was passed by Parliament. It made Non-Conformists who had taken the oath of allegiance exempt from penalties for non-attendance at the services of the Church of England.

Joseph Clarke, William's brother, worked as a physician in Grantham, after leaving his post as usher of the school, until his death in 1690. His only child Anne and her husband John Smith were the main beneficiaries of his will.

At the time of his death on 4 January 1692, Humphrey Babington was a very wealthy man, and owned property across Lincolnshire and Leicestershire. In his will written in 1686, he was a generous benefactor, leaving money to Arthur, Edward, Katherine and their mother Katherine Clarke, as well as many other relatives and God-children. None was left to Ann in Maryland, as she was not his niece, nor did he leave anything to Katherine's children by William Clarke, Joseph, Martha and John.

He also left money to have a hospital built in Barrow upon Soar, in memory of his uncle Theophilus Cave, a committed churchman, and made several bequests to the poor.[57] The hospital was for six aged, poor widowers or bachelors of good character from Barrow upon Soar and Quorndon, known as Theophilus Cave's bedesmen. Each man was issued with a blue cloak, faced with white, and was required to go to church each Sunday.

Barrow upon Soar Hospital

There is an interesting inscription on a memorial in the church
to Theophilus Cave, which reads:

Here in the Grave there lyes a Cave,
Well call a Cave a Grave,
If Cave be Grave and Grave be Cave
Then reader Judge I Crave
Whether doth Cave here lye in Grave
Or Grave here lye in Cave?
If Grave in Cave here buried lye
Then Grave where is thy Victorie?
Goe Reader and Report here Lyes a Cave
Who conquers Death and Buries his own Grave

Barrow upon Soar church

Edward Storer died in 1712 and was buried in Buckminster church. He appeared to have had a disagreement with his son Edward, since he wrote in his will 'Item I give unto my very undutifull son Edward Storer one shilling'. He left his brother's Universal Double Ring Dial to his family for safe keeping.[58] His grandson Bennet, a clergyman, was exonerated after a murder trial in London several years later.[59]

Edward Storer's grave in Buckminster Church

Katherine Vincent died shortly after giving her testimony to William Stukeley and was survived by one of her four children and her grandchildren.

Joseph Clarke was an apothecary in Loughborough. His first wife Rebekah died and he married Frances Key, the daughter of John Key, in 1690. His two children by Frances died in infancy and are buried with their mother. Joseph administered his uncle Babington's bequest in Barrow upon Soar and was also a generous benefactor when he died in 1721. He mentions his family connections in his will.[60] Martha Clarke married John Boyer, a maltster from Loughborough. They had six children, John, Robert, Catherine, Anne, Humphrey and

Joseph, a distiller in Nottingham. John Clarke married Anne and died in 1729 in Loughborough.

William Clarke their step-brother had three sons and a daughter. All three sons were surgeons and apothecaries, William the eldest studying at Utrecht medical school. One of the other sons, Ralph, and his wife Judith Welby had five daughters and six sons, several of whom died in childhood. One son Charles was an apothecary in Grantham and died in 1796 aged 49. He was a very wealthy man and left some of his money and property to his Welby cousins, and also to his sister Selina Muscutt, wife of George Muscutt, an apothecary in Grantham. Selina and George lived in Vine House on Vine Street, Grantham. This was another Storer family connection to Isaac Newton, since Judith Welby was the granddaughter of Isaac's grandfather's cousin Judith Newton of Gonerby and her husband William Welby.

Ann Skinner died in 1714 and was survived by six children and two step-children, as well as many grandchildren. She too was buried on her land at Prince Frederick. Many of her descendants were apothecaries, physicians and surgeons. Gravestones were visible on the land until the late twentieth century.

Martha Greenfield née Truman's grave in Maryland

Colonel Truman's grave in Maryland

On the site of Ann Skinner's land, where Arthur Storer is buried, there is now a planetarium called the Arthur Storer Planetarium which commemorates one of Lincolnshire most famous sons and North America's first astronomer known by name.

The Planetarium in the grounds of Calvert High School, Prince Frederick, Maryland.

ARTHUR STORER PLANETARIUM

ARTHUR STORER (c 1642-1686), THE FIRST ASTRONOMER IN THE AMERICAN COLONIES, CAME TO CALVERT COUNTY FROM LINCOLNSHIRE, ENGLAND HE WAS AMONG THE FIRST OBSERVERS TO SIGHT AND RECORD DATA DESCRIBING HALLEY'S COMET ON ITS RETURN IN 1682. HIS OBSERVATIONS WERE MADE FROM THIS TRACT OF LAND. ARTHUR STORER WAS A LIFELONG FRIEND AND COLLEAGUE OF SIR ISAAC NEWTON WHO QUOTED STORER'S DATA REPEATEDLY IN HIS GREAT SCIENTIFIC WORKS.

BOARD OF COUNTY COMMISSIONERS
CALVERT COUNTY HISTORICAL SOCIETY
AND
MARYLAND HISTORICAL SOCIETY

The Arthur Storer plaque in Prince Frederick Maryland

1645-1687

Family Tree

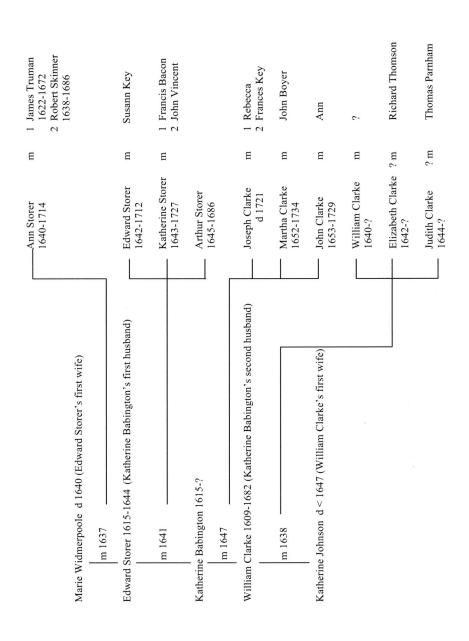

The Second Generation

Children of Ann Storer and James Truman
Elizabeth Truman 1660-1660
Martha Truman 1662-? m Thomas Greenfield
Mary Truman 1663-1699 m Thomas Halliday
Ann Truman 1664-1670
Elizabeth Truman 1666-1735 m Charles Greene
Boy Truman 1670-1670

Children of Ann Storer and Robert Skinner
Clarke Skinner c.1673-1714 m Ann
William Skinner c.1664-1738 m Elizabeth Mackall
Adderton Skinner c.1677-1756 m Rebecca Truman

Children of Robert Skinner and his first wife Alice Thomas
Nathaniel Skinner
Robert Skinner 1667-1713
Mary Skinner 1668-1696 m Joseph Letchworth

Children of Edward Storer and Susann Key
Elizabeth Storer
Edward Storer 1674-?
Mary Storer 1676-?
Arthur Storer 1678-1681
Susanna Storer 1679-1679
Susanna 1680-1681
Thomas Storer 1682-?
Anne Storer 1686-1750 m Ellis Key
Katherine Storer 1688-1766 m Edward Fane
Francis Storer 1691-1745 m Elizabeth

Children of Katherine Storer and Francis Bacon
William Bacon
Edward Bacon 1666-1708
Ann Bacon d 1677
Katherine Bacon d 1678

Children of Joseph Clarke and Frances Key
Infant Clarke died as infant
Infant Clarke died as infant

Children of Martha Clarke and John Boyer
John Boyer 1680-1746
Humphrey Boyer 1681-1716
Catherine Boyer 1684-?
Robert Boyer 1687-?
Joseph Boyer 1689-1730
Anne Boyer 1691-1692

Children of William Clarke
William Clarke 1670-?
Ann Clarke 1674-?
Mary Clarke 1677-?
Joseph Clarke 1679-?
Ralph Clarke 1685-1764 m Judith Welby

The Third Generation

Children of Martha Truman and Thomas Greenfield
Martha Greenfield
Jane Greenfield
Elizabeth Greenfield
James Greenfield

Anne Greenfield
Thomas Greenfield
Truman Greenfield

Children of Mary Truman and Thomas Halliday
James Halliday
Margery Halliday
Leonard Halliday

Children of Elizabeth Truman and Charles Greene
Ann Greene 1697-?
Mary Greene 1699-?
Charles Greene 1706-?

Children of Clarke Skinner and Ann
Ann Skinner 1700-1741

Children of William Skinner and Elizabeth Mackall
Henry Skinner 1714-1750 m Elizabeth Greenfield
Maryland Skinner 1716-1759
Dorcas Skinner 1724-?
Ann Skinner 1726-?
Major Skinner 1718-1785
Sarah Skinner 1722-1766
Robert Skinner 1720-?

Children of Adderton Skinner and Rebecca Truman
James Skinner
Joseph Skinner
Martha Skinner
William Skinner
Leonard Skinner
John Skinner

Mary Skinner

Children of Mary Skinner and Joseph Letchworth
Thomas Letchworth

Children of Anne Storer and Ellis Key
John Key 1723-1789
Ellis Key 1721-1721
Thomas Key 1724-?

Children of Francis Storer and Elizabeth
Bennet Storer 1726-1804
Catherine Storer 1728-?

Children of Edward Bacon
Francis Bacon 1700-?

Children of Ralph Clarke and Judith Welby
Mary Clarke 1723-?
Judith Clarke 1726-?
Joseph Clarke 1728-1730
Ralph Clarke 1729-1751
John Clarke 1730-?
Ann Clarke 1732-1732
Judith Clarke 1739-? m Thomas Shephard
Benjamin Clarke 1741-1742
Arthur Clarke 1743-1746
Charles Clarke 1747-1796
Selina Clarke m George Muscutt

References

1 *The Making of Grantham* (2011), edited by David Start and David Stocker, p. 54

2 Michael Honeybone, *The Book of Grantham* (1980), p. 77

3 Lincolnshire Archives, GB5/1, Hall Book of Grantham, fo. 89

4 Tracy Borman, *Witches: A Tale of Sorcery, Scandal and Education* (2013), p. 66

5 Nottinghamshire Archives, Transcription of the Archdeaconry Court of Nottingham 1565-1675, volume 3, p. 416

6 Ruth Crook and Barbara Jefferies, *The History of Little Gonerby and its School* (2008), pp 92-93

7 St Wulfram's church records

8 Lincolnshire Archives, INV/136/503

9 St Wulfram's church records

10 *House Of Commons Journal,* vol. 2 (1640-43), 27 June 1642, pp. 640-43; *HMC Portland Papers MSS,* I, p. 40; *The Private Journals of the Long Parliament 2 June to 17 September 1642,* edited by Vernon F Snow and Anne Steele Young, pp. 136-7, 142

11 John Manterfield, 'The Topographical Development of the Pre-Industrial Town of Grantham, Lincolnshire, 1535-1835', unpublished University of Exeter Ph D thesis, 1981

12 Lincolnshire Archives, GB5/1, Hall Book of Grantham, fos. 127, 137

13 Nottinghamshire Archives, Bishop's Transcripts, Wysall

14 Nottinghamshire Archives, Abstracts of Marriage Licenses, Archdeaconry Court 1577-1700

15 Nottinghamshire Archives, Bunny church registers

16 Buckminster Estate Office, Buckminster church registers

17 Lincolnshire Archives, GB5/1, Hall Book of Grantham, fo. 196

18 'January 1650: An Act For Subscribing the Engagement', *Acts and Ordinances of the Interregnum, 1642-1660* (1911), pp. 325-329; http://www.british-history.ac.uk/report.aspx?compid=56379

19 Bodleian Library, Oxford, Tanner MSS, Nalson 2, 11 June 1651, 27 June 1651, 30 July 1651, 13 August 1651

20 Lincolnshire Archives, GB5/1, Hall Book of Grantham, fo. 227

21 Lincolnshire Archives, GB5/1, Hall Book of Grantham, fo. 252

22 St Wulfram's church records

23 *Bunny: Images of the Past, Recording the History of a Nottinghamshire Village*, Bunny Millennium History Group, p. 23; *Oxford Dictionary of National Biography* (2004), online version, Babington, Humfrey; Lambeth Palace Library, MS Fil 1186

24 London, Royal Society Library, MS/142, http://www.newtonproject.sussex.ac.uk/

25 *The Correspondence of Isaac Newton*, edited by H.W. Turnbull and others, volume 7 (1977), p. 373: Storer's owing money to Newton

26 Fitzwilliam Museum, Cambridge, Fitzwilliam notebook by Isaac Newton

27 Royal Society Library, London, MS/142, available at http://www.newtonproject.sussex.ac.uk

28 Bodleian Library, Rawlinson MS 25, f.33, printed in R White, *Dukery Records* (Worksop, 1904), p. 227

29 Lincolnshire Archives, Hall Book of Grantham GB5/1, fo. 288

30 Lincolnshire Archives, Hall Book of Grantham GB5/1, fo. 292

31 Lincolnshire Archives, Hall Book of Grantham GB5/1, fo. 295

32 Lincolnshire Archives, Hall Book of Grantham GB5/1, fo. 303

33 Lincolnshire Archives, Hall Book of Grantham GB5/1, ff. 231, 301

34 Lincolnshire Archives, Hall Book of Grantham GB5/1, fo. 273

35 Lincolnshire Archives, Hall Book of Grantham GB5/1, fo. 303

36 Lincolnshire Archives, Hall Book of Grantham GB5/1, fo. 302

37 Lincolnshire Archives, Hall Book of Grantham GB5/1, fo. 315

38 Land records of Pr. Geo Co. Liber A, folio 97; Maryland State Archives, MSA SC 4341

39 *Calendar of State Papers Domestic: Charles II 1660-1,* p. 146

40 *Oxford Dictionary of National Biography* (2004), online version, Babington, Humfrey

41 Trinity College, Cambridge MS R.4.48c

42 Lincolnshire Archives, GB5/1, Hall Book of Grantham, fo. 365

43 Lincolnshire Archives, GB5/1, Hall Book of Grantham, fo. 382; Philip E. Hunt, *The Story of Melton Mowbray* (1957), p. 94

44 *The Visitation of Leicestershire 1619,* ed. J. Fetherston, Harleian Society volume XX (1870), p. 197

45 The National Archives, PROB 11/500/17

46 *Calendar of State Papers Domestic: Charles II 1663-4,* p. 338 and *1666-7,* pp. 126-7

47 *Calendar of State Papers Domestic: Charles II 1666-7,* p. 214

48 Dr Williams's Library, London, Baxter Correspondence, Volume IV, ff. 24-25; Maryland State Archives, MSA SC 4341

49 Lincolnshire Archives, Wills 1682/ii/465

50 Proceedings of the Council of Maryland, 1671-1681, edited by William Head Browne, *Maryland Historical Society* (Baltimore, 1896)

51 Md. Cal. Wills, Vol. I, p. 70; Wills, Liber 1, fol. 509

52 Cambridge University Library, MS Add. 3978/3, Storer's letter

53 Cambridge University Library, MS Add. 3978/4, Storer's letter

54 Cambridge University Library, MS Add. 3978/5, Storer's letter

55 Cambridge University Library, MS Add. 3978/6, Storer's letter

56 *The Principia*, translated by Andrew Motte (New York, 1995), pp. 417, 429

57 The National Archives, PROB 11/408, Babington's will

58 Leicestershire Record Office, Inven 251, Edward Storer's will

59 Bennet Storer court records http://www.oldbaileyonline.org

60 Leicestershire Record Office, Archdeaconry Court of Leicester 1660-1750, Joseph Clarke's will

Acknowledgements

My appreciation and thanks are firstly due to Helen Martin. Helen, who lives in Pennsylvania, visited Grantham in 2011 to give The Newton Lecture at The King's School, Grantham, followed by a talk on Sir Isaac Newton at the Gravity Fields Festival in 2012.

Since her visit in 2011, when we first met, we have had lengthy discussions in person, by telephone, email and Skype about Arthur Storer, his friendship with Isaac Newton and William Clarke. We decided that, with all the information that we had gathered, I would write a book for adults, whilst Helen, who is a National Board Certified Teacher, would write a childrens' book. Our aim is to make people aware of the life and work of Arthur Storer, one of Lincolnshire's and Maryland's famous sons.

My appreciation and thanks are also due to my husband Dr David Crook for proof reading, Dr John Manterfield for sharing his knowledge of Grantham and for proof reading and Prof Rob Iliffe for our email exchanges. To Rosemary Richards and Ameneh Enayat of the Gravity Fields Festival, Margaret Winn of Woolsthorpe Manor, Dr Peter Elmer of Exeter University, Michael J Schaffer, Courtney Finn of Grantham Civic Society, Paul Ambrose of Leicestershire and Rutland Archives, Marilyn Palmeri of the Pierpont Morgan Library, New York, Megan Craynon of Maryland State Archives, Jack Greene and The Rev Canon Michael Rock Covington.

Thank you also to Lincolnshire Archives, Nottinghamshire Archives, Leicestershire and Rutland Archives, The Bodleian Library, Oxford, Dr Williams' Library, London, The National

Archives, London, Cambridge University Library, Trinity College, Cambridge, The Pierpont Morgan Library, New York, Maryland Archives, Annapolis, The Parliamentary Archives, London, Lambeth Palace Archives, London, The National Portrait Gallery, London, The Royal Society, London, The King's School Grantham, South Kesteven District Council, Woolsthorpe Manor, Woolsthorpe, The Newton Project, The Whipple Museum, Cambridge, Moseley Old Hall, Staffordshire, The Usher Gallery, Lincoln, Lincoln Central Library, the staff of Cotgrave Post Office, Nottinghamshire, Getty Images, The Library of Congress, Washington and the staff of The Buckminster Estate.

Ruth Crook 2014

Parliamentarian helmets

Volunteer in Seventeenth century costume at Moseley Old Hall

Further reading

Lou Rose and Michael Marti, *Arthur Storer of Lincolnshire, England, and Calvert County, Maryland: Newton's friend, Star Gazer, and Forgotten Man of Science in Seventeenth-Century Maryland* (Maryland, 1984).
Comment: The authors make many assumptions. Some of the genealogy is inaccurate.

Peter Broughton, 'Arthur Storer of Maryland: His Astronomical Work and his Family ties with Newton', *Journal for the History of Astronomy*, 19 (1988) p. 77-96.
Comment: Details Storer's astronomical work and family. Some of the genealogy is inaccurate.

James D Trabue, 'Ann and Arthur Storer of Calvert County, Maryland, Friends of Sir Isaac Newton' *The American Genealogist* (January-April 2004) p. 13-27.
Comment: Concerns Plantagenet connections. Some of the genealogy is inaccurate.

Helen E Martin, 'The Friendship of Arthur Storer and Sir Isaac Newton', *The Calvert Historian,* The Calvert County (Maryland) Historical Society 2012.
Comment: Written during our joint research.

Helen E Martin and Bonita Evans-Gondo, 'On the Shoulders of Sir Isaac Newton and Arthur Storer', *The Science Teacher*, Vol. 80, No. 2, Feb 2013.
Comment: Written during our joint research.

Grantham Civic Society

Aims:

- Ensure that the town's remaining fine or historically interesting buildings are preserved for future generations

- Monitor the upkeep of green spaces and make our own contribution, where appropriate, to their provision

- Keep a watching brief on new developments and make constructive comments on their likely impact on the town's environment

- Encourage sustainable development and refurbishment

- Encourage a vibrant, attractive, economically viable environment in which to work, live and for leisure

  Hence our motto: **Preserve the good in the old; encourage the good in the new.**

Grantham Civic Trust, as it was known then, was formed in the early 1960s by local people who had become greatly concerned about extensive demolition and poorly planned development in the town.

Over the years we have maintained a good relationship with the planning authorities but kept a critical eye on developments. Our Planning Sub-committee reviews significant planning applications which affect historic buildings and other major developments. Our comments are passed to the Planning Dept of South Kesteven District Council which makes decisions on proposed developments.

The Society has played a role over the years in achieving listed status for important buildings under threat of demolition. The district council often consults us on major plans and we have attended many meetings, conferences and visits to other towns over the years as part of the planning process.

The Society has been involved in tree planting schemes and has carried out small scale landscaping projects. The Society was instrumental in the formation of the **Rivercare** group which continues to be successful in maintaining and improving the River Witham and keeping the parkland through which the river flows clear of rubbish.

Grantham Civic Society has established a triennial **Townscape Awards scheme** to encourage good new building, conservation and landscape improvements. Over the years the Society has had a beneficial and steady influence on the development of the town. However, the ravages of the 60s and later decades are still all too evident and have left us with a town centre below the standard one expects from a community the size of Grantham. The challenges of traffic, pollution, and parking all remain perhaps in ways different from past times. The District Council is more concerned than ever before not to make the mistakes of the past but there will always be a need for an independent point of view. The Society provides this as long as we represent a large body of informed people.

In recent years the Society has started to erect commemorative **Blue plaques** to celebrate the lives of famous people who have lived in or passed through the town. **Information signboards** have been commissioned which graphically explain the history of parts of the town. We believe these things inform townspeople and visitors alike and help to create a sense of place which makes people prouder of their town and perhaps more interested in looking after it.

The Society publishes newsletters, holds speaker meetings, social events and organises visits to places of interest. We host other Civic societies and give guided tours of the town. Our visits elsewhere will often be hosted by another Civic society. Nationally, we are part of **Civic Voice** which is the national charity for the civic movement which plays a vital part in seeking to influence government policy towards good design and responsible development.

Grantham Civic Society is delighted to be involved with this new book by Ruth Crook after the success of her 2013 book: *The History of Vine House and Vine Street, Grantham.*

www.granthamcivicsociety.co.uk

List of Subscribers

Patricia Aveyard
Prof Graham Baker
Tiffany Biezack
Roger Blakeman
John & Mary Bomphrey
Sam Branson
Robert Brownlow
Brenda Butters
Marilyn & Malcom Campbell
Russell Cann
Graham Cook
Richard & Andrea Coppin
James A Cousins
David & Marion Ellis
David Feld
Courtney Finn
Roland Fisher
Harry Green
David & Elaine Green
Stephen Hallam
Harry Hamer
Harlaxton College
Sebastian Hearmon
Dr Basil & Sylvia Hiley
Keith & Jil Hiley
David Hindmarch
David Hill
Robert Hirst
Elizabeth Holmes
Michael Honeybone

Linda Houtby
Chanti Humphrey
Cressie Humphrey
Freddie Humphrey
Esther Humphrey
Charlie Humphrey
Prof Robert Iliffe
Dr Terry Jones
Simon Jowitt
Dr Gordon Kingsley
Gerald P LaBelle
Bijal Ladva
Mary Losure
Dr John & Barbara Manterfield
Luke & Stacia Manterfield
Helen E Martin
Howard Matthews
Irvin Metcalf
Adam Metcalf
Ian Mihill
Louise Mumford
Alison Paxton
Christopher Pettit
John & Patricia Pinchbeck
Vera Quick
Peter Reichelt
Martin Rodell
Paul Ross
Prof David Sleight
Henry Sleight

Isobel Sleight
Marie Spillane
Valerie Taylor
The King's School
Jeff Thompson
Colin Thorpe
Richard Tuxworth
Stephen Vogt
Christopher Watkin
Janice Webb
Simon Webb
John Williams
Elizabeth Wilson
Linda Wootten